T0195500

Mandatory Workplace Safety and Health Programs

Implementation, Effectiveness, and Benefit-Cost Trade-Offs

Tom LaTourrette, John Mendeloff

Sponsored by the Commonwealth of Pennsylvania

CENTER FOR HEALTH AND SAFETY
IN THE WORKPLACE

This study was sponsored by the Commonwealth of Pennsylvania and was conducted in the RAND Center for Health and Safety in the Workplace.

Library of Congress Cataloging-in-Publication Data

LaTourrette, Tom, 1963-
    Mandaory workplace safety and health programs : implementation, effectiveness, and benefit-cost trade-offs / Tom LaTourrette, John Mendeloff.
        p. cm.
    Includes bibliographical references.
    ISBN 978-0-8330-4557-7 (pbk. : alk. paper)
    1. United States. Occupational Safety and Health Administration. 2. Occupational health services—Standards—United States. 3. Medical policy—United States. 4. Industrial safety—United States. I. Mendeloff, John M. II. Rand Corporation. III. Title.
    [DNLM: 1. United States. Occupational Safety and Health Administration. 2. Safety Management—standards—United States. 3. Accidents, Occupational—prevention & control—United States. 4. Health Policy—United States. 5. Occupational Diseases—prevention & control—United States. 6. Occupational Health Services—standards—United States. WA 485 L361w 2008]

RC968.L38 2008
363.11—dc22
                                                                                    2008034475

The RAND Corporation is a nonprofit research organization providing objective analysis and effective solutions that address the challenges facing the public and private sectors around the world. RAND's publications do not necessarily reflect the opinions of its research clients and sponsors.

RAND® is a registered trademark.

Published 2008 by the RAND Corporation
1776 Main Street, P.O. Box 2138, Santa Monica, CA 90407-2138
1200 South Hayes Street, Arlington, VA 22202-5050
4570 Fifth Avenue, Suite 600, Pittsburgh, PA 15213-2665
RAND URL: http://www.rand.org/
To order RAND documents or to obtain additional information, contact
Distribution Services: Telephone: (310) 451-7002;
Fax: (310) 451-6915; Email: order@rand.org

Preface

In 1998, the Occupational Safety and Health Administration (OSHA) began a rulemaking process for a standard that would require all employers to establish a workplace safety and health program. Employer opposition to the standard and a focus on higher-priority issues led OSHA to abandon that effort. Based on recent interest in learning about a mandatory standard, this report critically examines several key issues that emerged from the rulemaking process. It presents recommendations for analyses and other steps that could help inform decisions about whether to implement a safety and health program standard. This report should be of interest to OSHA, state labor departments, and researchers interested in occupational safety and health.

The RAND Center for Health and Safety in the Workplace

This work was sponsored by the Commonwealth of Pennsylvania and was conducted in the RAND Center for Health and Safety in the Workplace. The Center for Health and Safety in the Workplace is dedicated to reducing workplace injuries and illnesses. The Center provides objective, innovative, cross-cutting research to improve understanding of the complex network of issues that affect occupational safety, health, and workers' compensation. Its vision is to become the nation's leader in improving workers' health and safety policy.

The Center is housed at the RAND Corporation, an international nonprofit research organization with a reputation for rigorous and objective analysis on the leading policy issues of our time. It draws upon the expertise within three RAND research units:

- RAND Institute for Civil Justice, a national leader in research on workers' compensation
- RAND Health, the most trusted source of objective health policy research in the world
- RAND Infrastructure, Safety, and Environment, a national leader in research on occupational safety.

The Center's work is supported by funds from federal, state, and private sources.

For additional information about the Center, please contact:

John Mendeloff, Director
Center for Health & Safety in the Workplace
RAND Corporation
4570 Fifth Avenue, Suite 600

Pittsburgh, PA 15213-2665
John_Mendeloff@rand.org
(412) 683-2300, x4532
(412) 683-2800 fax

Contents

Preface iii

Tables vii

Summary ix

Acknowledgments xiii

Abbreviations xv

CHAPTER ONE

Introduction 1

CHAPTER TWO

The Proposed Safety and Health Program Standard 3

Plan Components 3

Management Leadership and Employee Participation 3

Hazard Identification and Assessment 4

Hazard Prevention and Control 4

Information and Training 4

Evaluation of Program Effectiveness 4

Clarity and Enforceability 4

Existing Safety and Health Program Standards 6

CHAPTER THREE

Evidence on the Effectiveness of Safety and Health Programs in Preventing Injuries 7

What Is the Evidence? 8

Case Reports from Individual Firms 8

The Experience of Recognized High-Performing Facilities 8

Comparing Participants and Nonparticipants in State Programs 9

Injury Rate Changes in States That Have Adopted Safety and Health Program Requirements 10

Conclusions About Evidence on the Effects of Safety and Health Programs on Injuries 16

CHAPTER FOUR

Benefits and Costs of the Proposed Safety and Health Program Rule 17

Industry Baseline 17

Cost 18

Effectiveness 19

Monetizing Benefits 20

Comparison of Benefit and Cost Estimates 21

Cost-of-Illness Approach...21
Willingness-to-Pay Approach... 22
Summary... 23

CHAPTER FIVE
Recommendations for Further Analysis.......................................25
Effectiveness 25
Separate the Effect of Safety and Health Programs from Other Factors That Influence Injury
    Rates...25
Examine in More Depth the Experience from Existing Programs 25
Implementation and Enforcement 26
Benefits and Costs 27

Bibliography..29

Tables

3.1. Features of Three Studies on the Impact of State Safety and Health Program
      Standards 11
3.2. Changes in State Total Recordable Rates for Private Industry Five Years After
      Adoption of Mandatory Safety and Health Program Requirements 13
4.1. OSHA's Estimate of the Costs of the Proposed Safety and Health Program Standard ... 19
4.2. Break-Even Effectiveness Estimates 22

Summary

In 1998, the Occupational Safety and Health Administration (OSHA) began to develop a standard that would have required all workplaces to establish a "safety and health program." A safety and health program uses management tools that address general behaviors and procedures to reduce the risk of occupational injuries and illnesses. Safety and health programs typically have four main components:

- management commitment and employee involvement
- worksite analysis
- hazard prevention and control
- safety and health training.

Several states already had regulations requiring safety and health programs. Others—and OSHA itself—provide incentives for employers to voluntarily adopt safety and health programs. OSHA argued that establishments that had safety and health programs had lower rates of injuries and illnesses. OSHA chose to abandon its rulemaking effort, partly because its proposal for an ergonomics standard drew most of the agency's standard-setting resources away from other areas and partly due to opposition from the business community. Critics of the proposed standard charged that the requirements were vague and left too much discretion to inspectors, that the evidence for the effectiveness of mandates for safety and health programs was unconvincing, and that the cost to employers of implementing such programs was very high and greatly underestimated by OSHA.

Interest remains at both the federal and state levels in finding ways to increase the prevalence of safety and health programs. As a contribution to the discussion of this issue, this report examines the evidence on effectiveness, costs, and benefits that was cited by different parties during the 1998 rulemaking and in more recent studies.

This report addresses the key question of whether mandatory safety and health programs are effective. Assessing the likely impact of a safety and health program standard requires two separate steps. The first is to estimate the effect on baseline injuries that would result from adopting a certain set of practices. The second is to estimate the extent to which employers will actually adopt those practices.[1] It is important to note that our analysis is concerned with mandatory safety and health programs. We have not carefully evaluated the evidence on the effectiveness of voluntary safety and health programs. There is certainly evidence to suggest

---

[1] Note that, for the purpose of establishing the "feasibility" of a new standard, OSHA generally assumes that there will be full compliance. Thus, the second step does not enter into that analysis. In contrast, if we are trying to estimate the expected benefits and costs of a new standard, it is necessary to take compliance into account.

that firms that voluntarily and conscientiously administer safety and health programs achieve reductions in injuries and illnesses.

We reviewed a limited set of studies and found that, although they mostly suggest that mandatory safety and health programs reduce injuries and illnesses, there are methodological and confounding factors that render their conclusions uncertain. Thus, these studies do not permit confidence in the effectiveness of mandatory safety and health programs.

As a result, we developed a sensitivity analysis that examines how effective a mandatory safety and health program would have to be to generate benefits that exceed its costs. Using our interpretation of OSHA's data and analyses and improved measures of benefits, we found that reductions in injuries of less than 10 percent could probably generate a positive net benefit. Of course, this conclusion rests on a number of assumptions that are open to challenge. It would be helpful to carry out a more thorough analysis, and, consequently, we suggest several strategies for learning more about the program effects and costs.

Our study also recognized that a mandatory safety and health program standard raises a variety of legal and enforcement issues. Employers expressed concern that requirements would be too vague to tell them clearly what had to be done and would leave too much discretion to compliance officers. They also worried about "double jeopardy," e.g., being cited for both a hazard-specific violation and a safety program violation for not having identified the hazard in its required surveys of the workplace. To gain perspective, we suggest that future studies try to learn from OSHA directors and employers in states with mandatory programs about whether these concerns have materialized in a significant way.

Based on our analysis, we make some recommendations for further research and other steps to clarify the effectiveness, implementation issues, and benefit-cost trade-offs of safety and health programs.

Effectiveness

A number of different types of analyses could help encourage better understanding of the effectiveness of mandatory safety and health programs:

- Examine whether establishments cited for safety and health program violations had higher injury rates prior to the citation than did other firms that were inspected but not cited.
- Examine whether establishments cited for safety and health program violations showed improvement in their injury rates subsequent to the citation, compared to similar establishments that did not get cited.
- Look for associations between safety and health program violations and intermediate metrics of effectiveness, such as measures of management commitment and worker engagement, changes in hazard identification rates, and changes in violation rates of other OSHA standards.
- Conduct more research to understand how changes in state workers' compensation programs could affect workers' reporting of injuries and illnesses.

Implementation and Enforcement

To better clarify the issues and impediments related to the implementation and enforcement of a safety and health program standard, it would be valuable to address the following questions:

- How frequently is the safety and health program standard cited relative to other standards, how often are such violations cited as "serious," and which elements of a safety and health program standard are most commonly cited?
- What are the states' enforcement policies, and is there any relationship between these and the evidence about the effectiveness of the state programs?
- What type of training do inspectors receive to judge compliance and enforce the standard? Are there specific training tools or approaches that have been particularly successful?
- What sorts of communication efforts and other special assistance do states provide to employers prior to and during the early phases of implementation?
- What type of feedback have states received from employers regarding implementation and enforcement, and how have states responded to feedback?

Benefits and Costs

An updated and improved analysis of the benefits and costs of a safety and health program standard would benefit from efforts to

- clarify the current industry baseline in terms of workers and establishments that have compliant safety and health programs
- consider the impact of safety and health programs on all injury types rather than just lost-workday injuries
- gather improved data on program costs from interviews, site visits, surveys, and stakeholder input.

Acknowledgments

This report benefited from valuable discussions with Robert Burt (OSHA), Jasbinder Singh (Policy Planning and Evaluation, Inc.), John Howard (National Institute for Occupational Safety and Health), Christine Baker (California Commission on Health and Safety and Workers' Compensation), Len Welsh (California Division of Occupational Safety and Health), and Dave Bellusci (California Workers' Compensation Insurance Rating Bureau). We also gratefully acknowledge insightful peer reviews from David Weil (Boston University) and Brian Gifford (RAND).

Abbreviations

OMB	Office of Management and Budget
OSHA	Occupational Safety and Health Administration
SBAR	Small Business Advocacy Review
SBREFA	Small Business Regulatory Enforcement Fairness Act of 1996
SHARP	Safety and Health Achievement Recognition Program
SIC	Standard Industrial Classification
VPP	Voluntary Protection Programs

CHAPTER ONE

Introduction

In 2006, more than 4 million people in the United States suffered work-related injuries and illnesses. The Occupational Safety and Health Administration (OSHA) was created to promote the safety and health of the country's workers by "setting and enforcing standards; providing training, outreach and education; establishing partnerships; and encouraging continual process improvement in workplace safety and health" (OSHA, 2008). Since passage of the Occupational Safety and Health Act (P.L. 91-596) and data collection by OSHA in 1972, the reported incidence of lost-workday occupational injury and illness in the United States was essentially flat, except for cyclical changes, until the early 1990s, when it began to decline substantially. While many factors influence occupational injury and illness rates, a significant one is regulation of workplace safety and enforcement of safety and health standards. Along with OSHA, 21 states have taken advantage of a provision of the Occupational Safety and Health Act that allows states to adopt their own standards and enforcement policies for private-sector employees, providing that they meet or exceed federal OSHA standards. Such states are known as "state-plan" states.

This report focuses on a particular workplace safety and health promotion initiative known as a *safety and health program*. A safety and health program is a workplace intervention that uses management tools to reduce the risk of occupational injuries and illnesses. (A more complete description is provided in the next chapter.) OSHA's main emphasis has been mandatory, government-enforced compliance with standards targeting specific workplace hazards (Hatch et al., 1978). Over time, however, it has shown interest in more generic standards that address important management practices. By focusing on safe behaviors and procedures, safety and health programs are designed to complement more conventional interventions that target specific hazards (e.g., machine guards). In 1982, OSHA instituted its Voluntary Protection Programs (VPP), which provide incentives for employers to implement workplace safety and health programs. OSHA also released guidelines for use by employers in designing and implementing a safety and health program (OSHA, 1989). In addition, starting at least as early as 1973, state-plan states began mandating safety and health programs or promoting voluntary efforts to implement them.

In the early 1990s, labor unions and their supporters tried but failed to pass legislation that would have required firms to establish joint labor-management safety committees. A major role of these committees would have been to oversee safety and health program activities. Following that defeat, OSHA began work to develop a standard that would require all workplaces to establish a safety and health program but that avoided the more controversial safety-committee requirement. Based on its experience with the VPP and state safety and health programs, OSHA had become convinced that safety and health programs were effective in reducing workplace injuries. By the end of 1998, OSHA had prepared a draft standard (OSHA, 1998a),[1] an initial regulatory flexibility analysis (focused on small-business impacts; OSHA, 1998b), a preliminary economic analysis (OSHA, 1998d), and a preliminary effectiveness analysis (OSHA, 1998c). In addition, it had consulted with small-business representatives under the aegis of a Small Business Advocacy Review (SBAR) panel comprised of staff from OSHA, the U.S. Small Business Administration, and the Office of Management and Budget (OMB), as required by the Small Business Regulatory Enforcement Fairness Act of 1996 (SBREFA); (P.L. 104-121). A hearing on the draft standard was held by the Republican-led U.S. House of Representatives Committee on Small Business in July 1999 (House Committee on Small Business, 1999).

OSHA's rulemaking attempt met with strong opposition from some business representatives and lawmakers. Opponents were dissatisfied on several fronts. First, they challenged OSHA's claim that mandatory safety and health programs were effective in reducing injury rates. They also contended that OSHA had greatly underestimated the costs of implementing and maintaining a safety and health program. Finally, they expressed concerns that the proposed standard was too vague to be effectively and fairly enforced. Shortly thereafter, OSHA abandoned plans to formally propose a standard. This step resulted not only from business opposition but also from OSHA's decision to focus its resources in the last year of the Clinton administration on promulgation of a standard on ergonomic hazards, which was a higher priority for organized labor than the safety and health program standard.

We were asked by several state agencies to look into what was known about the desirability of mandatory safety and health programs. In light of this interest, we undertook a study to examine the key issues that were raised in the 1998–1999 rulemaking process. We first describe the key elements of a safety and health program and summarize key questions and concerns that were raised by opponents regarding the implementation and enforcement of the proposed standard (Chapter Two). We then draw from documents related to the rulemaking process, as well as the small amount of academic literature on the subject, to critically evaluate evidence pertaining to the effectiveness of safety and health programs (Chapter Three). This is followed by a comparison of the benefits and costs of the proposed safety and health program standard using both OSHA's original approach and 1998 data and an alternate approach and updated data (Chapter Four). We conclude with recommendations for analyses and other steps to help resolve key open questions related to the clarity, enforcement, effectiveness, benefits, and costs of mandatory safety and health programs (Chapter Five). The objective of the recommendations is to provide new information to better inform decisions about whether to implement a federal or state safety and health program standard.

---

[1] Note that OSHA's draft proposed standard was not published as a notice of proposed rulemaking and so was not formally a proposed standard.

CHAPTER TWO

The Proposed Safety and Health Program Standard

Safety and health programs complement regulations targeting specific hazards by focusing on general management and organizational systems, with the intention of promoting safety. In 1989, OSHA published guidelines for designing safety and health programs (OSHA, 1989). According to this guidance, an effective program includes provisions for the systematic identification, evaluation, and prevention or control of general workplace hazards, specific job hazards, and potential hazards that may arise from foreseeable conditions (OSHA, 1989). This chapter describes the components of a safety and health program as defined in the proposed standard and then discusses some key concerns related to implementation and enforcement of the standard, as raised in the rulemaking process.

Plan Components

The draft proposed standard (OSHA, 1998a) identifies five main components of a safety and health program.

Management Leadership and Employee Participation

Employers must

> [e]stablish the program responsibilities of managers, supervisors, and employees for safety and health in the workplace and hold them accountable for carrying out those responsibilities; [p]rovide managers, supervisors, and employees with access to relevant information, training, and resources they need to carry out their responsibilities; [i]dentify at least one manager, supervisor, or employee to receive and respond to reports about workplace safety and health conditions and, where appropriate, to initiate corrective action. . . . Regularly communicate with employees about workplace safety and health matters; [p]rovide employees with access to information relevant to the program; [p]rovide ways for employees to become involved in hazard identification and assessment, prioritizing hazards, training, and program evaluation; [e]stablish a way for employees to report job-related fatalities, injuries, illnesses, incidents, and hazards promptly and to make recommendations about appropriate ways to control those hazards; and [p]rovide prompt responses to such reports and recommendations. (OSHA, 1998a)

Hazard Identification and Assessment

"The employer must systematically identify and assess hazards to which employees are exposed and assess compliance with the General Duty Clause and OSHA standards" (OSHA, 1998a).[1] This includes "conduct[ing] inspections; review[ing] safety and health information; [evaluating] new equipment, materials, and processes for hazards before they are introduced into the workplace; and assess[ing] the severity of identified hazards and ranking those that cannot be corrected immediately according to their severity" (OSHA, 1998a).

Hazard Prevention and Control

Employers must "systematically comply with the hazard prevention and control requirements of the General Duty Clause and OSHA standards. If it is not possible for the employer to comply immediately, the employer must develop a plan for coming into compliance as promptly as possible, which includes setting priorities and deadlines and tracking progress in controlling hazards" (OSHA, 1998a).

Information and Training

Employers must ensure that "[e]ach employee is provided with information and training in the safety and health program; and [that] each employee exposed to a hazard is provided with information and training in that hazard," including how to recognize it, what is being done to control it, what protective measures the employee must follow to prevent or minimize exposure to it, and the provisions of applicable standards (OSHA, 1998a).

Evaluation of Program Effectiveness

Employers must "evaluate the safety and health program to ensure that it is effective and appropriate to workplace conditions" (OSHA, 1998a).

Clarity and Enforceability

It is evident that the proposed safety and health program standard does not apply to hazards resulting from specific equipment, materials, or processes. Rather, it defines broad steps that are intended to reduce the risk of injury or illness resulting from any workplace hazard. This aspect was central to a key concern raised during the rulemaking process: Some stakeholders felt that the proposed standard was too vague to be effectively and fairly enforced. The SBAR Panel report (1998) and discussion during the hearing (House Committee on Small Business, 1999) cited examples of unclear intentions and vague language regarding how often employers must conduct hazard inspections (whenever "appropriate to safety and health conditions at the workplace"), how often employers must provide employee training and evaluations ("as often as necessary"), and what constitutes adequate employee training and sufficient employee involvement.

Concerns about clarity and enforceability portend the anticipated difficulty of implementing a mandatory safety and health program standard and highlight the distinction between

---

[1]   Section 5(a)(1) of the Occupational Safety and Health Act (P.L. 91-596), commonly referred to as the "General Duty Clause," specifies that "[e]ach employer shall furnish to each of his employees employment and a place of employment which are free from recognized hazards that are causing or are likely to cause death or serious physical harm to his employees."

voluntary and mandatory programs. Small-business representatives in the SBREFA process and James Talent, chair of the House Committee on Small Business, did not oppose incentives for implementing voluntary programs (SBAR Panel, 1998; House Committee on Small Business, 1999). However, given the vagueness of the proposed standard, they felt that employers could never be confident about what they had to do to be in compliance with a mandatory program. They felt that the vagueness of the standard placed too much discretionary authority with OSHA safety and health inspectors. As Talent claimed, "In essence, you make the inspector the policeman, the judge, the jury."

At the hearing, OSHA director Charles Jeffress responded that OSHA had already started working with its consultant staff[2] to provide them with clear and consistent training on what constitutes an effective safety and health program, how to help an employer set up an effective program, and how to evaluate a program. Jeffress acknowledged that some aspects of the proposed rule could be interpreted as vague and noted that OSHA was actively revising the rule to minimize uncertainty.

At the same time, Jeffress contended that laws and regulations are often based on a "reasonable-person" test (i.e., What would a reasonable person do?) and that this proposed rule was not unusual in this sense. He also emphasized that OSHA purposely designed the proposed rule to be flexible and nonprescriptive rather than to use a one-size-fits-all approach. He stated in his testimony that OSHA "has not specified every action a business must take to comply. Nor should it" (House Committee on Small Business, 1999). We note that several states and Canadian provinces have had mandatory safety and health program standards— many for 10 years or more. Statistics from state labor departments demonstrate that these standards have been enforced, at least in some states,[3] indicating that a functional safety and health program standard is possible.

A number of participants in the SBREFA process expressed concern about OSHA's enforcement policy. Questions arose about whether a violation of a specific OSHA standard would also constitute a violation of the proposed safety and health program standard. Its draft enforcement policy (as summarized in SBAR Panel, 1998) stated that "OSHA generally will not use this safety and health program rule to penalize employers twice for the same offense." The policy indicates that, in cases in which penalties are assessed for a violation of a particular standard or the General Duty Clause, penalties would be assessed under the safety and health program standard only if the employer has also systematically failed to identify and control significant hazards. It also indicates that, if there were no violations of underlying standards, OSHA would not penalize employers under the safety and health program rule (SBAR Panel, 1998). It therefore appears that a violation of a particular standard or the General Duty Clause is a prerequisite for an employer to be assessed a penalty under the safety and health program standard.

---

[2] OSHA provides free, nonpunitive, safety and health consultations to all small businesses (< 250 employees) in the country to help them identify and correct safety and health hazards.

[3] For example, the safety and health program standard is the most commonly cited violation in California and Washington and the 13th most commonly cited in Hawaii and North Carolina.

Existing Safety and Health Program Standards

At least 14 states already have some form of mandatory safety and health program standard. In its regulatory analysis, OSHA indicated that as many as 25 states may have such a rule (OSHA, 1998b). Regardless of the precise number, the existence of state standards raises the question of what authority these standards would have if a federal standard were promulgated. Existing standards in state-plan states would be allowed to continue as long as they were judged by OSHA to be at least as effective as the federal rule (OSHA, 1998b). In response to stakeholder input, OSHA included a grandfather clause in the proposed standard. However, this clause exempts employers in these states only if that state's program "satisfies the basic obligation for each core element" of the federal standard (OSHA, 1998a). This is not a grandfather clause in the conventional sense, as it exempts only employers that are already substantially in compliance. Thus, as envisioned in 1999, all states would need to come into compliance with the proposed standard.

The fact that several states have existing mandatory safety and health program standards, and that a large number of employers have implemented voluntary programs, also has important implications for assessing the effectiveness and costs of safety and health programs. As discussed in the next chapter, evidence for program effectiveness is sometimes clouded by difficulties in controlling for the existing programs in comparison groups. The high and uncertain prevalence of existing programs also complicates benefit and cost estimates for the safety and health program standard: Establishments that already have programs will neither incur the costs nor realize the benefits of implementing one. OSHA tried to estimate the prevalence of safety and health programs that complied with the proposed standard in 1998 (i.e., the existing industry baseline) from information about mandatory state programs and a large survey of employers that it conducted in 1993 (OSHA, 1998d). OSHA estimated that 23 percent of establishments had compliant programs and that 51 percent of all employees covered by the proposed rule worked in establishments that contained the core elements (PPE, 1999). Given these large numbers, the requirements for employers with existing programs will be important for assessing the effects of new federal or state rules.

Evidence on the Effectiveness of Safety and Health Programs in Preventing Injuries

One of the major questions to arise from the 1998 rulemaking process was whether safety and health programs had been shown to be effective in reducing workplace injuries. In this chapter, we examine the evidence presented during that process, as well as some newer data, to assess the extent to which the effectiveness of mandatory safety and health programs has been demonstrated.

To help understand the potential effectiveness of a mandatory safety and health program standard, it is useful to consider how adoption of a safety and health program standard is expected to prevent injuries and illnesses. OSHA never explicitly articulated the process by which it envisioned this occurring, but it is reasonable to infer that the objective was to produce changes in the workplace that would lead to the elimination of hazards and the substitution of safer behavior for less safe behavior. Both outcomes can result from changes in processes that lower risk, and a mandatory standard increases awareness of the elements of a good safety and health program.

Increasing awareness of the elements of a good safety and health program is an important first step toward preventing injuries, but to understand the extent to which a mandatory standard would ultimately reduce injuries, we would ideally answer a series of questions:

1. To what extent does having a mandatory standard increase the utilization of the elements in a safety and health program?
2. Does stricter enforcement increase utilization?
3. Does more extensive informational outreach increase utilization?
4. To what extent does utilization prevent workplace injuries?

Unfortunately, the ability to answer those questions now, as in 1998, is limited by the paucity of well-designed evaluations and useful data. Our evaluation therefore examines the combined effect of these steps and addresses the overarching question of whether mandatory safety and health programs prevent injuries and illnesses.

Given OSHA's stated enforcement policy, one issue that looms over this discussion is whether safety and health programs that firms would establish under this policy would be just "paper programs," undertaken for compliance but not backed by any real managerial commitment. Our ability to directly measure this commitment is very limited, but we can try to determine whether other mandatory programs have shown signs of success. In the remainder of this chapter, we examine the studies and data that have been adduced as evidence regarding the effectiveness of safety and health programs in preventing injuries and illnesses.

What Is the Evidence?

There are several types of evidence that have been used in the debate about the effectiveness of safety and health programs. These include

- case reports from individual firms that workplaces that have introduced safety and health programs experienced large drops in injuries
- reports that, of the firms recognized as having the lowest injury rates (e.g., those in OSHA's VPP), almost all have safety and health programs
- comparisons of injury rates within a state—between firms in voluntary programs and firms that are not and between firms subject to a mandatory program and those that are not
- comparisons of changes in injury rates in states with and without mandatory safety and health programs.

We address each of these types of evidence in turn.

Case Reports from Individual Firms

In his 1999 testimony, Charles Jeffress, OSHA's director, presented some examples and testimonials from individual firms that have adopted safety and health programs voluntarily and reported that their injury rates have decreased (House Committee on Small Business, 1999). Such evidence is not very germane to predicting the likely effectiveness of a mandatory program covering all or most firms. There are several possible biases:

- The firm may have adopted it after a bad year, in which case the apparent improvement may reflect no more than the increased probability that the injury rate in the following years would be closer to its long-term average (i.e., "regression-to-the-mean" bias).
- Firms that choose to voluntarily adopt safety and health programs may possess above-average "management commitment," which has been shown to be important and cannot be assumed in a mandatory program (i.e., sample-selection bias).
- People are more likely to report and publish what they perceive to be successes. Unsuccessful efforts are less likely to be reported (i.e., publication bias).

The Experience of Recognized High-Performing Facilities

As mentioned previously, OSHA has programs to recognize firms that have shown high levels of safety performance. The VPP initiative, established in 1982 and now recognizing about 1,400 facilities, is the most demanding in terms of both low injury rates and a demonstrated commitment to a strong safety and health program. A similar but less demanding program, the Safety and Health Achievement Recognition Program (SHARP), grew out of OSHA's Consultation Project and provides services to firms (except large ones) that request advice on compliance. Charles Jeffress testified in 1999 that experience with these programs shows that the high-achieving firms relied on safety and health programs to achieve those results (House Committee on Small Business, 1999).

It may be true that safety and health programs did help those firms achieve reductions in injuries and illnesses; however, the relevance of that issue to the likely effectiveness of a mandatory safety and health program requirement is uncertain. The chief reason is the sample-selection bias, noted earlier. The improvements for the committed facilities that volunteer for SHARP and, especially, VPP will be greater than those experienced under a mandatory program. The key question is whether a reluctant firm, or one with limited resources, can be pushed to carry out a program with enough vigor to make a difference. In addition, there are questions about whether the firms with a strong stake in a good safety reputation take steps to suppress reportable incidents. In these programs, OSHA has required significant employee involvement, which may prevent such practices. No studies have looked closely at this issue, as far as we are aware.

Comparing Participants and Nonparticipants in State Programs

In its analysis (OSHA 1998c), OSHA first identified "studies or data relevant to the effectiveness of safety and health programs" in California, Colorado, Maine, Massachusetts, Michigan, North Dakota, Oregon, and Texas. OSHA acknowledged that the data from the states varied greatly and did not always directly address the likely impact of a standard such as the one OSHA was considering. Indeed, the California and Michigan studies discussed injury prevention but had no specific focus on or insights about safety and health programs.

The Maine Top 200 Program had required larger firms that wanted to escape regular inspections to adopt key features of a safety and health program. The Maine program won awards for its innovative approach, but the two studies on the program's impact on injuries do not allow firm conclusions to be drawn about its effectiveness (see Mendeloff, 1995, and Stanley, 2000).

In 1982, Oregon had required that establishments with relatively high injury rates create a safety and health program. However, this initiative was, according to OSHA, "seen as ineffective due to a variety of reasons including uneven enforcement," which in 1991 led to a requirement for joint labor-management safety committees. The committees must carry out activities similar to those required by OSHA's draft safety and health program. A team at the University of Oregon produced an unpublished evaluation (Hecker, Gwartney, and Barlow, 1995), which reported that the "Oregon's safety committee regulations seem to have had some fairly strong effects on OR-OSHA enforcement outcomes. . . . Whether and how these enforcement outcomes are linked to injury rates in Oregon is not established."

The Texas Extra-Hazardous Employer Program, which began in 1991, targeted firms whose injury frequency rates were substantially higher than those of other firms in their industries. This program required participants to institute safety and health programs that resembled those in OSHA's draft proposal. OSHA presented data on the changes in injury rates among participants relative to changes among all Texas employers. Firms that completed the program had reductions of 49 to 77 percent between the 12 months prior to entry and 12 months after entry. One major problem with this analysis was that the changes among participants were provided only for firms that *completed* the program. The data show that, for each year from 1992 through 1994, fewer than 30 percent of the firms identified for participation in the program actually completed it. In 1995, the percentage rose to just over 40 percent.[1]

---

[1]   A second potential problem is regression-to-the-mean bias due to choosing firms on the basis of their high rates in the previous year. If firms join a program at least in part because their injury rate jumped to an unusually high level, we would expect that, in the next year, the rate would decline regardless of any interventions, simply because there is a random element in the occurrence of injuries.

In evaluating programs, a frequent question is whether to look only at the firms that complete the program or at all that were supposed to complete the program. Looking only at the former can determine whether those that stuck it out did, in fact, show an improvement. But then we face the problem of a self-selected, voluntary group, which may not be representative of all intended participants. If we want to know the effect of the program on all of those who were initially required to participate, it is appropriate to include all the dropouts and see what happened to injuries for the entire group. Even then, however, evaluators face the problem of comparing the changes for that group with those of a comparable set of nonparticipants. Thus, comparing the changes in the group to changes among all Texas employers would probably not be appropriate. Instead, the evaluation should try to find a similar group of "extra-hazardous" employers that were not involved with this program. That may not be feasible in Texas if the policy included all extra-hazardous firms.

Colorado, Massachusetts, and North Dakota all had voluntary safety and health programs that were linked to discounts on firms' workers' compensation premiums. The data reported by OSHA did indicate that the voluntary participants experienced greater decreases in injury rates than nonparticipants. As already noted, however, voluntary programs are notoriously difficult to evaluate, because firms that choose to join probably differ from those that do not. In both Colorado and Massachusetts, the participants had unusually high injury rates when they entered the program. Although there are methods to try to control for observable differences that can affect participation, unobserved differences (for example, in management motivation) remain a significant obstacle to isolating the effect of participation. As a result, the findings from these programs are probably of limited relevance for assessing the likely impact of a mandatory standard.

None of these state studies provides much evidence for the effectiveness of mandatory safety and health programs. Of the eight studies examined by OSHA, two did not address safety and health programs (California and Michigan), one was inconclusive (Maine), four likely suffered from sample-selection bias (Texas, Colorado, Massachusetts, and North Dakota), and the evaluations of the Oregon program did not claim to demonstrate reduced injury rates.

Injury Rate Changes in States That Have Adopted Safety and Health Program Requirements

A number of studies of varying degrees of sophistication have looked at the injury rates of states with and without safety and health programs and tried to make inferences about the effectiveness of the programs. In this section, we consider three analyses of these comparisons:[2] OSHA's analysis prepared in 1998 as part of the regulatory impact analysis (OSHA, 1998c), an analysis by Conway and Svenson (1998), and a study by Smitha, Kirk, et al. (2001).[3] Table 3.1 compares some central features of these studies.

It is interesting that the studies disagreed about which states should be considered to have safety programs. However, the most important point is that the studies differ in what they were testing. Conway and Svenson (1998) tried to explain injury rate changes from 1994 to 1996. They found that the rates in states with mandatory health and safety programs, whether under state OSHA or workers' compensation programs, did not decline more during those two

---

[2]   In addition, we consider the critique of OSHA's analysis commissioned by the U.S. Small Business Administration and carried out by Policy Planning and Evaluation, Inc. (PPE, 1999).

[3]   Smitha, Oestenstad, and Brown published a more descriptive study about the same issue in *Professional Safety* (see Smitha, Oestenstad, and Brown, 2001).

Table 3.1
Features of Three Studies on the Impact of State Safety and Health Program Standards

Study Parameter	Study		
	Smitha, Kirk, et al. (2001)	**Conway and Svenson (1998)**	**OSHA (1998c)**
State and year standard adopted			
Safety and health program requirement	AK (1985–1995) MT (1993) CA (1991) NE (1994) HI (1984) NV (1994) LA (1983) NH (1994) MN (1991) WA (1973)	AK MN CA NV FL NC HI OR MI WA	AK[a] LA CA[a] MN (1995) FL OR HI[a] WA[a]
Safety committee requirement	CT (1995) NE (1994) FL (1993) NV (1994) OR (1991) NH (1994) MN (1995) WA (1973?) MT (1993)		
Sector	Manufacturing	All	All but construction, agriculture, and mining
Rate	Lost workday	Lost workday and total recordable[b]	Total recordable[b]
Sample years	1992–1997	1994–1996	3–5 years after adoption of state health program
Comparison group	States without state health program laws	States without mandatory state health programs	U.S. rate
Controls	Many	None	None
Conclusions	States with state health programs had lower rates	Rates in states with state health programs did not decline relative to comparison group	Rates in states with state health programs for 5 yrs or more[b] declined 18% relative to U.S. rate

[a] The states that had state health programs for at least five years at the time of the study were Alaska, California, Hawaii, and Washington.

[b] Total recordable rate is the total number of injuries and illnesses per 100 full-time-equivalent workers (BLS, 2007).

years than the rates in states without such requirements. Of the three studies, Conway and Svenson (1998) is probably the least relevant to predictions about the impact of a new federal standard. Of the eight states identified by OSHA as having mandatory safety and health programs, only three had begun in the 1990s. Therefore, although not irrelevant for those three, the changes from 1994 to 1996 can provide only an incomplete picture of the impact of safety and health programs.

OSHA (1998c) tried to estimate the changes in total recordable injury rates that occurred in the three- to five-year period following the introduction of a safety and health program. In concept, the OSHA study's test is probably the most appropriate for an evaluation. If safety and health programs are effective, we probably should expect to see an effect emerge over the first three to five years after the program begins.

Smitha, Kirk, et al. (2001) tried to estimate the average effect of having a safety and health program (and other requirements) on the *level* of injury rates in manufacturing from 1992 through 1997. The study concluded that the presence of safety and health program laws was associated with reduced lost-workday rates (as were laws requiring safety committees and laws requiring loss-control activities by insurers). Given that many factors may affect injury rates, testing the assumption that safety and health programs are important depends on being able to control for these other effects. The Smitha, Kirk, et al. (2001) study, alone among the three, does try to do that.

In the following sections, we focus on the OSHA (1998c) and Smitha, Kirk, et al. (2001) studies.

The OSHA Analysis of State Program Effectiveness. In a draft of its effectiveness analysis (OSHA, 1998c), OSHA stated,

> To determine the impact of OSHA's proposed program on injury and illness incidence rates, this analysis compares the changes in these rates for eight pre-selected states to those in the US as a whole. As the eight states under review have already implemented safety and health programs similar to that proposed by OSHA, the results of this analysis may shed some insight into the possible impact of the proposed standard.

The eight states that OSHA included were Alaska, California, Florida, Hawaii, Louisiana, Minnesota, Oregon, and Washington. OSHA also stated that it was excluding other states with safety and health programs because their programs did not cover all or almost all of the workers in the state.

One point to note is that all of the eight but Florida and Louisiana are state-plan states, in which enforcement is carried out by state OSHA programs, not federal OSHA. The significance of that fact is that the OSHA programs in those six states can directly enforce the state's safety and health program requirements as part of their regular enforcement activities. In Florida and Louisiana, in contrast, the safety and health program requirement is not enforced by federal OSHA because it is not a federal standard. In those two states, if there is to be enforcement of the safety and health program requirement, it has to be carried out by other authorities, perhaps the state labor department or the workers' compensation agency.[4]

OSHA made several choices in its design for evaluating these programs. First, it chose to use the total recordable injury and illness rate as the measure of impact. OSHA did not specify why it chose this measure rather than the equally plausible lost-workday rate, which includes only cases with either lost workdays or restricted work activity.

Second, OSHA's measure of impact was the change in that rate from the year after adoption of the safety and health program until two to four years later, compared to the change in the overall U.S. rate for those years. Using these multiyear averages might obscure the impact of the program if it tends to become more effective over time, but the cumulative reduction may be more indicative of the effect of the program.[5] Also, it would have been better to compare the rates of the states with safety and health programs to the states without the programs (rather than to the entire United States); the practice used could impart a downward bias to the estimate of impact if states with safety and health programs did indeed do better.

---

[4] In fact, Conway and Svenson note in their article that Florida had "limited State enforcement" of its safety and health program requirement (Conway and Svenson, 1998, Table 4, note 19). In that case, Florida should not be used to predict the effects of stronger enforcement.

[5] For example, suppose that, using the year after the program was established as the baseline, we see a 1-percent reduction in the second year, a 2-percent reduction in the third year, a 3-percent reduction in the fourth year, and a 4-percent reduction in the fifth year. The average reduction, compared to the baseline year, would be 2.5 percent (10 percent divided by four years).

Third, OSHA chose to analyze the changes for each industry sector separately, which is a reasonable way to get more insights.[6] However, it did not present the results for each state separately. That could be a potentially serious shortcoming, because we would not know whether the results for the group of states were really driven by only one or two of them.

OSHA found that safety and health programs had been in operation long enough to provide five years of "post-" data in only four of the eight states (Alaska, California, Hawaii, and Washington), so it made comparisons for all eight states for a three-year follow-up period and used a five-year follow-up period for the other four. In Table 3.2, we present rate changes for seven of the states, adding data for more recent years to include the same number of follow-up years for all. We excluded Florida because of its limited enforcement of its safety and health program.

For four of the seven states that OSHA had identified as having programs that covered almost all workers, the decreases following adoption exceeded the U.S. drop by a sizable margin. However, the rate changes in Oregon and Alaska showed almost no difference from the change in the national rates, and the rate in Minnesota dropped much less. Across the seven states, on average, the state rates dropped by 5 percent more over the five-year period than did the U.S. rate. OSHA had reported an average 18-percent drop for the four states it had selected (OSHA, 1998c). A very large limitation of this analysis, as OSHA acknowledged, is that it fails to consider other factors that may have influenced changes in rates.

Table 3.2
Changes in State Total Recordable Rates for Private Industry Five Years After Adoption of Mandatory Safety and Health Program Requirements

Characteristic	States Identified by OSHA							States Identified by Others	
	CA	HI	MN	OR	AK	LA	WA	NE	NC
Year state health program adopted	1991	1982	1991	1991	1985	1983	1973	1993	1993
Total recordable rate in next year	9.3	10.6	8.6	9.1	10.2	7.9	12.3[a]	10.2	7.5
Total recordable rate 4 years later	6.6	9.8	8.3	7.8	11.1	7.4	10.5	8.5	6.1
% change	−29	−8	−3	−14	9	−6	−14	−17	−13
% change in U.S. rate for same period	−17	9	−17	−17	11	8	−5	−20	−20

NOTE: Total recordable rate is total number of injuries and illnesses per 100 full-time-equivalent workers. The total recordable injury and illness rates for states and the United States are reported by the Bureau of Labor Statistics annual Survey of Occupational Injuries and Illnesses (BLS, 2007). As an example of the injury rate change calculations, for California, the table compares the rate in 1992 with the rate in 1996.

[a] State-level injury rate data were not available prior to 1976. Therefore, the rate of 12.3 in Washington is the rate for 1976, not 1974, and 10.5 is the rate for 1980.

---

[6] It also reasonably excluded agriculture, mining, and construction, which are frequently excluded or treated differently by safety and health programs.

In its report, the SBAR Panel raised the question of how safety and health programs could be effective if states with them had higher injury rates than states without them. The panel observed that

> OSHA bases its benefits analysis in part on its estimates of the success of State programs that contain requirements similar to those in the proposed rule in reducing the incidence of job-related illness and injuries. The Panel recommends that OSHA include more details of this analysis to justify this preliminary conclusion in the face of other evidence that seems contradictory. For example, Washington state has enforced comprehensive health and safety regulations for two decades, but over the past decade, the rate of injuries and illnesses reported in that state remains between 20 and 40 percent higher than the national average. Similarly, over the past decade, Minnesota's incidence rates fell below the national average, but in the two years after implementation of the state's health and safety regulations, the state's rates have exceeded the national average. The Panel also notes that illness and injury incidence rates vary from year to year and that a similar pattern of year-to-year variations arises in states with health and safety programs and in states without them. The Panel recommends that OSHA more clearly display the basis for its preliminary conclusions that state health and safety programs are effective in reducing job-related injuries and illnesses. (SBAR Panel, 1998)

The answer to this question is that states differ in more than whether they have a safety and health program. One example is the workers' compensation waiting period. Several studies have shown that the number of injuries reported to the Bureau of Labor Statistics is higher in states with shorter waiting periods before lost wages are compensated and with lower levels of benefits. Our own review suggests that, on average, states with a three-day waiting period have reported lost-workday rates that are about 15- to 20-percent higher than the rates reported in states with a seven-day waiting period. Six of the eight states that OSHA identified as having extensive and long-running safety programs, including Washington, had three-day waiting periods. This indicates that caution should be used when comparing absolute injury and illness rates among different states.

Close observers of changes in workers' compensation laws have pointed to many changes (beyond waiting periods and maximum benefit levels) that may affect the probability that workers will file claims. For example, many states have limited coverage when the injury was an aggravation of a preexisting condition; others now require more "objective" medical evidence, especially for musculoskeletal injuries; and restrictions have grown on compensation for psychological injuries (Burton, 2008). Thomason and Burton (2001) estimated that changes in Oregon statutes from 1987 to 1995 had reduced the number of claims by between 12 percent and 28 percent by 1996.

We have not taken into account the possible impact of differences in "reporting culture." It seems likely that some states inculcated stronger traditions of reporting than other states did. The fact that surveys find that large numbers of workers report that they had work injuries that they did not report to workers' compensation agencies suggests that there is considerable room for fluctuations in response to reporting incentives. Louisiana, which is notorious for poverty, obesity, and corruption, has the lowest reported lost-workday injury rate. In fact, when we look at fatality rates in construction, there is evidence that the correlation between the lost-workday rates and fatality rates across states is actually negative. Washington, which has one of the highest injury rates, appears to have the lowest fatality rate. Thus, there may be a positive relationship between poor safety programs and poor reporting of nonfatal injuries and illnesses. All these factors need to be considered when examining studies such as that of Smitha, Kirk, et al. (2001).

Smitha, Kirk, et al. (2001) Study of Workplace Safety Laws and Occupational Injury Rates. Smitha, Kirk, et al. (2001) make a serious effort to control for a number of other factors that affect the level of lost-workday injuries in manufacturing. Like other studies, it uses the Bureau of Labor Statistics Survey of Occupational Injuries and Illnesses (BLS, 2007) for its measure of injury rates. Unlike the others, it does not focus primarily on rate changes over time. Rather, Smitha, Kirk, et al. attempt to account for the relative contribution of several potential influences on the level of injury rates. By focusing on the level of the injury rate, it implicitly assumes that the effects of these potential influences, including a safety and health program, are long-term.

The study controls for the following factors:

- Standard Industrial Classification (SIC) (industry composition)
- year effects from 1992 through 1997
- the annual number of OSHA inspections per 1,000 manufacturing workers
- the size of OSHA fines per manufacturing worker
- the number of OSHA consultation program visits per 1,000 manufacturing workers
- average firm size in industry in 1992
- whether the state had a state plan to run enforcement
- the workers' compensation maximum payment
- the workers' compensation waiting period
- the percentage of high-school graduates
- state unemployment rate
- percentage of manufacturing workforce that was unionized
- a series of variables for the percentage of workers in different age groups.

The sample consisted of data from 21 industries and 42 states over about four years, for a total of 3,286 observations. (Many states did not have Bureau of Labor Statistics data for all two-digit SICs for all the years.) For each state, the authors tried to determine the following:

- the percentage of the workforce affected by requirements for mandatory implementation of employer health and safety programs
- the percentage of the workforce affected by mandatory safety committee requirements
- whether the state had provisions requiring that workers' compensation insurers provide loss-control services
- whether the state had laws that called for targeting high-hazard employers.

The analysis included all four of these variables together; thus, it tried to estimate the effect of each requirement, holding the others constant. In Table 3.1, we showed which states were counted under each of the first two requirements. Of the 10 states listed as having safety and health program requirements, and of the nine states listed as having safety committee requirements, six are on both lists, and five of those show adoption of both in the same year. As a result, the coefficient generated for the safety and health program variable will be heavily determined by the four states that do not also have safety committee requirements. These are

Alaska, California, Hawaii, and Louisiana. As noted, three of the four had the largest changes in rates following the establishment of the safety program requirements.

Smitha, Kirk, et al., found that both safety committee requirements and safety program requirements were associated with reductions in lost-workday injuries in manufacturing during this period, i.e., the coefficients were statistically significant.[7] The authors also found, as expected, that longer workers' compensation waiting periods were strongly associated with lower reported injury rates. However, they also found, contrary to expectations, that higher maximum benefits under workers' compensation were also very strongly associated with *lower* reported injury rates. Because other studies have found the opposite (e.g., Ohsfeldt and Morrisey, 1997), this finding raises questions about the appropriateness of the model used in the study.

These points—the heavy reliance on only a few states and the unexpected finding for workers' compensation benefits—suggest that some caution is needed before accepting the claim that these findings show that safety and health programs do cause a drop in lost-workday injury rates in manufacturing.

Conclusions About Evidence on the Effects of Safety and Health Programs on Injuries

Even with the benefit of additional years, it is difficult to arrive at a clear conclusion about whether safety and health programs have reduced injury and illness rates. It seems very likely that, in workplaces that adopt the safety program and actively implement it, some injury reduction is likely to occur over the subsequent few years. But we do not know how many workplaces are likely to fit this description or how large the gains there would be.

In light of the failure to find unambiguous and consistent effects among states that have adopted safety and health programs, it seems plausible that any reduction in injury rates in workplaces that do not currently have programs is likely to fall well below OSHA's lower-bound estimate of 20 percent.[8] If the standard affects a smaller percentage of employees in a state, the overall impact on the state rate will be correspondingly smaller.

At the end of this report, we return to the issue of whether new research might be able to provide better answers concerning the effects of safety and health programs. Another perspective can come from asking a different question: How large would the effects of safety and health programs on reducing injuries have to be for them to be worthwhile? Insights on this topic depend on an assessment of values attached to the preventive effects and to the costs of achieving them. We now turn to that topic.

---

[7] Depending on the model used, the estimated effect of safety programs ranged from 4 percent to 12 percent, and the estimated effect for safety committee requirements ranged from 8 percent to 13 percent.

[8] Despite presenting evidence that implementing safety and health programs leads to an 18-percent average state decrease in injury rates relative to the national average (see Table 3.1), OSHA assumed in its initial regulatory analysis that a new federal standard would have an effectiveness of between 20 percent and 40 percent (OSHA, 1998b, 1998d). Note that the 18-percent value came from OSHA's estimate of the actual effect of implemented programs and thus reflects both program effects and the degree of compliance. OSHA's regulatory analyses are supposed to estimate both costs and effects under the assumption of full compliance, which will lead to a higher-than-observed effectiveness.

CHAPTER FOUR

Benefits and Costs of the Proposed Safety and Health Program Rule

A central element of OSHA's draft regulatory analysis is a benefit-cost analysis that compares the cost of implementing and operating a federal safety and health program standard to the cost savings, or benefit, resulting from the reduction in injuries and illnesses anticipated as a result of the standard. In this chapter, we use data presented in OSHA's analysis, along with some alternative values, to compare the expected benefits and costs of the proposed standard. In addition to illustrating the relationship among effectiveness, cost, and benefit, this chapter highlights areas that might warrant reexamination or further analysis. It is important to note that OSHA's economic analysis was ongoing at the time the rulemaking effort was abandoned, and the available data were preliminary (see OSHA, 1998d).

First, we present what is known about the industry baseline and how more detailed estimates could be derived. Next, we present OSHA's estimates of the costs to employers of implementing and operating a safety and health program. Then, we present two approaches for valuing the benefits of safety and health programs in terms of reduced injuries and illnesses. To estimate the total benefit, we assess the injuries expected to be prevented under a new standard. Finally, we present a break-even analysis to determine how effective a safety and health program standard would need to be for its benefit to equal its cost.

Industry Baseline

Many states currently have some type of mandatory safety and health program, and many employers have implemented voluntary programs (OSHA, 1998b, 1998d). This means that many workers already participate in a safety and health program and therefore would incur no cost and see no benefit from a new federal standard. As noted previously, OSHA estimated in 1998 that 23 percent of establishments had compliant programs (PPE, 1999). It also indicated that the total number of affected establishments is 5.9 million. Based on this total, the 23-percent industry baseline equals 1.4 million establishments. For the purposes of regulatory benefit-cost analysis, OSHA assumed that all employers that did not already have programs would come into full compliance with the proposed standard when it became active (OSHA, 1998d). Thus, the costs and benefits of a new standard would apply to the 4.5 million establishments not included in this baseline. In reality, not all employers will comply, so the costs and benefits presented here overestimate the actual values that would result from the proposed standard.

OSHA also estimated that 51 percent of all employees covered by the proposed rule worked in establishments that had compliant programs (PPE, 1999). Effectiveness and benefits were estimated in terms of worker injury and illness cases, so the proposed rule would apply to the number of injury and illness cases sustained by workers in establishments without compliant programs. OSHA did not explicitly state its estimate of the total number of cases in its initial regulatory and preliminary economic analyses. However, it did present anticipated reductions in the numbers of cases expected for different levels of effectiveness. Based on these results, we calculated that OSHA used a total of 2.9 million to 3.2 million lost-workday injuries and 2,100 to 2,300 fatalities. The number of lost-workday injuries is consistent with the annual number of lost-workday cases from 1994 to 1998 (2.8 million to 3 million; BLS, 2007). Using 3 million lost-workday cases, a baseline of 51 percent is 1.5 million cases, and a new standard would apply to the 1.5 million cases not included in this baseline.

Note that the total number of fatalities that OSHA used is much smaller than the annual number of occupational fatalities in the same period (5,000 to 5,600; BLS, undated). Transportation incidents and assaults and violent acts accounted for 60 percent of occupational fatalities in 1998 (BLS, undated); excluding these cases reduces the total to 2,000 to 2,240, which is about the same as OSHA's estimate. Thus, we infer that OSHA excluded fatalities from transportation incidents and assaults and violent acts from the total affected population in its analysis. Excluding these fatalities is probably warranted, as safety and health programs are not likely to have a substantial impact on these hazards.

Cost

In its preliminary economic analysis, OSHA estimated the costs of implementing and maintaining the proposed standard (OSHA, 1998d). Cost estimates were broken into several separate elements, and costs for each element were estimated based on a number of inputs, including time requirements, labor costs, turnover rates, current injury and illness rates, current industry baseline rates of compliance, and number of hazards that would be identified and need to be controlled. Costs were calculated separately at the three-digit SIC code (~300 industries) and for each of seven establishment size classes. These separate cost estimates were aggregated to generate a national total. OSHA indicated that the costs take into account the industry baseline and thus represent the incremental cost of a new standard (OSHA, 1998d). The individual cost elements and total annualized costs for the first 10 years of the program are shown in Table 4.1.

The average annual cost per establishment is the total cost divided by the 4.5 million establishments not in the industry baseline, or $1,100 to $1,500. About $500 of this is for the basic program elements (i.e., net of hazard-control costs). For the workplaces affected by a new standard, the figures will be smaller still because the firms with existing voluntary programs are disproportionately large ones. The average employer cost for hourly compensation in 1996, the year for which these figures were calculated, was just over $18. So $500 would cover about 28 hours per year. Some small-business representatives claimed that the costs would be 10 to 20 times larger than OSHA's estimates (SBAR Panel, 1998). The SBAR Panel concurred that the OSHA estimates were low, and made several recommendations for improving the accuracy and transparency of the cost estimates.[1]

---

[1]   The SBAR Panel recommended more clearly stating assumptions, providing more information on the time frame over which costs and benefits would be incurred, distinguishing initial cost layouts from recurring costs, and seeking input from

Table 4.1
OSHA's Estimate of the Costs of the Proposed Safety and Health Program Standard

Cost Element	Annual Cost ($ millions)
Management leadership (training individuals involved in the program and their program-related responsibilities)	357
Employee involvement	45
Hazard identification and assessment	379
Employee information and training	1,012
Program evaluation and program updates	422
Multi-employer worksites	111
Basic program subtotal	2,326
Hazard control[a]	2,563–4,378
Total	4,889–6,704

SOURCE: OSHA (1998d).

[a] OSHA's hazard-control cost estimates vary with program effectiveness; the cost range shown is for 20- to 40-percent effectiveness.

The most costly element in Table 4.1 is controlling or eliminating hazards identified through the safety and health program. OSHA assumed that greater effectiveness of a safety and health program would be achieved (at least in part) by greater reductions in hazards, which would, in turn, require higher costs. As a result, OSHA's hazard-control cost estimates increase as effectiveness increases. OSHA indicates that a 20-percent reduction in injuries from the baseline would require annualized hazard-control costs of almost $2.6 billion. Then, an additional 20-percent reduction (a 40-percent reduction from the baseline) would add another $1.8 billion in annualized costs (for a total annualized hazard-control cost of almost $4.4 billion). This is a surprising result because we generally assume that the marginal costs of prevention increase as we prevent more and more injuries. In other words, the costs of achieving a reduction from 20 percent to 40 percent would be considerably larger than the cost required for the first 20-percent reduction. In contrast, OSHA's figures imply that the costs would be almost 30 percent lower.

Effectiveness

The studies summarized in Table 3.1 in Chapter Three show that the effectiveness of mandatory safety and health programs has been estimated in terms of the reduction in lost-workday injuries and total recordable injury and illness cases. Given the general nature of safety and health programs, they are expected to reduce most types of injuries and illnesses to an equal extent. Thus, either metric is probably valid. Safety and health programs are also expected to

---

individual firms on the reasonableness of assumptions and estimates. The panel also recommended that OSHA more clearly include hazard-control costs as part of the full effects of the standard (SBAR Panel, 1998).

affect levels of fatal injuries, though no studies have examined this. As noted earlier, in its regulatory analysis, OSHA assumed that safety and health programs do not affect fatalities caused by transportation incidents or assaults and violent acts.

For the purposes of a benefit-cost analysis, the method used to estimate benefits must account for both the inherent effectiveness of a mandatory standard and employer compliance with the standard. Because not all employers will come into full compliance, the overall effectiveness will be lower than the inherent effectiveness. It might be possible to anticipate compliance with a federal standard based on inspection and violation data from mandatory state programs. However, we are not aware of any studies that have attempted this.

Given the great uncertainty about both anticipated compliance rates and the average effectiveness for firms that do comply (see Chapter Three), we did not attempt to estimate an effectiveness value. Rather, we have conducted a break-even analysis, discussed in the remainder of this chapter, to determine how effective the standard would need to be for its benefits to equal its costs.

Monetizing Benefits

OMB's and most economists' preferred approach for monetizing benefits for regulatory benefit-cost analysis is to identify individuals' willingness to pay for the anticipated change in conditions (OMB, 2003; Tolley, Kenkel, and Fabian, 1994). Willingness to pay can be estimated from revealed preferences methods, which exploit existing markets to identify the value that people place on a particular good or benefit. One common approach to valuing reductions in risk at work is to use hedonic wage studies to estimate the pay differential for taking on more hazardous jobs (see, e.g., Viscusi and Aldy, 2003). This approach has been adopted by the Environmental Protection Agency and is advocated by OMB. The literature suggests that workplace injuries and fatalities may be valued at roughly $70,000 and $6 million, respectively; however, these figures reflect a wide range of estimates.

As an alternative to the willingness-to-pay method, benefits that are realized in the form of improved safety and health are sometimes monetized using an approach known as the *cost-of-illness* or *human-capital* method (see, e.g., Blincoe et al., 2002). This method focuses primarily on medical costs and lost productivity resulting from injuries and illnesses and does not account for the reduction in quality of life that willingness-to-pay approaches attempt to capture. Willingness-to-pay estimates for reductions in risk (which include a quality-of-life component) have been shown to be far higher than medical or lost-productivity costs (see, e.g., Tolley, Kenkel, and Fabian, 1994); thus, cost-of-injury estimates are considered to greatly underestimate the value of avoided injuries and illnesses.

At the time of the 1998 rulemaking process, OSHA had a policy of not placing a monetary value on human life or on the pain and suffering experienced by injured or sick workers (OSHA, 1998d). Consistent with this policy, OSHA used a cost-of-illness approach to value the benefits of the proposed safety and health program rule (OSHA, 1998b, 1998d). OSHA's approach captured the injured worker's lost productivity, medical expenses, legal expenses, cost of administration of workers' compensation insurance, and indirect costs to employers (those unrelated to workers' compensation insurance). The resulting benefit of $12,700 per injury was applied to each avoided lost-workday injury (OSHA, 1998d).

In its regulatory analyses, OSHA now uses willingness-to-pay measures for both fatal and nonfatal outcomes.[2] By using a willingness-to-pay approach, the benefit calculated for a given level of effectiveness will be considerably higher than the value at which OSHA arrived using the cost-of-illness approach. We illustrate this point with an example later in this report.

Comparison of Benefit and Cost Estimates

In this section, we examine the relationship among effectiveness, benefits, and costs to explore the benefit-cost trade-offs of the proposed standard under different conditions. Specifically, we ask how effective at reducing injuries and illnesses a safety and health program standard would need to be for its benefit to equal its cost. This is known as a *break-even* or *threshold* analysis and is useful in situations in which benefits or costs are difficult to quantify or effectiveness is uncertain (OMB, 2003).

In general, the benefits of a proposed rule increase with the rule's effectiveness, while costs are nominally fixed. In the case of OSHA's proposed safety and health program standard, OSHA assumes that one cost component, the hazard-control cost, also increases with effectiveness. However, the remaining program cost elements are independent of effectiveness, so the net benefit still increases with effectiveness. The break-even point is found by adjusting the effectiveness until the net benefit equals zero.

We estimate the break-even effectiveness two ways, using different data and assumptions. In the first, we use OSHA's cost estimates and cost-of-illness benefit values to compare results from 1998. In the second, we use a willingness-to-pay benefit valuation approach and update the parameters to account for changes since 1998.

Cost-of-Illness Approach
Our cost-of-illness calculation uses data presented by OSHA in 1998 to estimate the break-even effectiveness for costs, benefits, and injury rates that existed in 1996. Parameter values and results are shown in Table 4.2. The basic program cost was taken from OSHA's baseline-adjusted cost estimates (see Table 4.1). For our analysis, we needed to estimate how much hazard-control costs increase for each extra percent of injuries that are prevented. As noted, we found OSHA's estimated relationship between hazard-control costs and effectiveness implausible because it implied that the marginal cost of preventing injuries declined as more and more injuries were prevented. We adopted a somewhat more plausible assumption that the marginal costs were constant.

Based on OSHA's estimate of the hazard-control costs to get to a 20-percent reduction ($2.56 billion), the cost per percent redution would be $128 million. If we took the cost to go all the way from doing nothing to a 40-percent reduction ($4.38 billion), the cost per percent reduction would be $110 million. Here, we use OSHA's cost estimate at the intermediate effectiveness value of a 30-percent reduction ($3.46 billion). This gives a cost of $115 million per percent reduction, or a hazard-control cost equal to $11.53 billion times effectiveness.

To estimate the benefit, we started with the total number of lost-workday injuries and fatalities inferred from OSHA's preliminary economic analysis. We then used OSHA's industry worker baseline estimate of 51 percent to calculate baseline-corrected numbers of cases.

---

[2] Personal communication with Robert Burt, director, Office of Regulatory Analysis, OSHA.

Table 4.2
Break-Even Effectiveness Estimates

Benefit or Cost	Benefit-Valuation Method and Relevant Year	
	Cost of Illness (1996)	Willingness to Pay (2006)
Basic program cost[a,b]	$2.33 billion	$3.30 billion
Additional (hazard-control) cost of preventing 1% of baseline-corrected injuries and fatalities[b,c]	$115 million	$164 million
Total affected lost-workday injuries[d]	3.0 million	2.1 million
Total affected fatalities[d]	2,100	2,500
Industry worker baseline[e]	51%	51%
Baseline-corrected lost-workday injuries	1.47 million	1.03 million
Baseline-corrected fatalities	1,030	1,230
Value of prevented lost-workday injury[f]	$12,700	$70,000
Value of prevented fatality[f]	Not considered	$6 million
Value of preventing 1% of baseline-corrected injuries and fatalities	$187 million (0.01 × 1.47 million injuries × $12,700)	$794 million (0.01 × [1.03 million injuries × $70,000 + 1,230 deaths × $6 million])
Break-even effectiveness[g]	33%	5.2%

[a] Data from OSHA (1998d) (see Table 4.1 herein).

[b] Willingness-to-pay values are increased by 42 percent over the cost-of-illness value to account for increased labor costs between 1996 and 2006 (see BLS, undated).

[c] The hazard-control cost is $11.53 billion times effectiveness.

[d] Cost-of-illness values from OSHA (1998d). Willingness-to-pay values from the Bureau of Labor Statistics (BLS, undated, 2007). Fatality totals exclude fatalities caused by transportation incidents and assaults and violent acts.

[e] Data from PPE (1999).

[f] Cost-of-illness values from OSHA (1998d). Willingness-to-pay values from Viscusi and Aldy (2003).

[g] Determined by solving for the level of effectiveness (E), where benefit equals cost. For example, for the 1996 analysis, cost = $2.33 billion + $115 million(100E). Benefit = $187 milllion(100E). Therefore, cost equals benefit when 2,330 = 7,200(E), or E = 33 percent.

Avoided lost-workday injuries and fatalities were determined by applying the effectiveness estimate to the baseline-corrected case numbers. To value the benefit, we used OSHA's estimate of $12,700 for each avoided lost-workday injury. Consistent with OSHA's policy at the time, we did not place a value on avoided fatalities. Using this valuation approach, we found that the effectiveness would have to be at least 33 percent for the benefits to be at least as large as the costs (see Table 4.2).

Willingness-to-Pay Approach

Because the cost-of-illness method does not account for the pain and suffering associated with injuries, we recalculated the break-even effectiveness using a willingness-to-pay valuation approach that attempts to incorporate pain and suffering. We also updated the costs and injury and fatality numbers to reflect current values (see third column in Table 4.2).

Costs, including the hazard-control cost, were updated to 2006 values by increasing OSHA's cost estimates (see Table 4.1) by 42 percent to account for the increase in labor costs since 1996 (BLS, undated). The total number of lost-workday injuries and fatalities in 2006 (the most recently available) were taken from BLS (2007, 2008). Having no new information about the current industry worker baseline, we used the same 51 percent estimated by OSHA in 1998 to calculate the baseline-corrected number of cases. We used benefit values of $70,000 per avoided lost-workday injury and $6 million per avoided fatality, which are the approximate midpoints of the ranges presented in a meta-analysis by Viscusi and Aldy (2003).

For a given number of avoided cases, the willingness-to-pay approach gives a benefit that is nearly six times greater than the benefit that OSHA estimates with the cost-of-illness method. This difference reflects the fact that individuals' willingness to pay to avoid injuries or death is much greater than the medical and lost-productivity costs included in the cost-of-illness approach. Thus, even after accounting for the higher costs and decreased numbers of injuries and fatalities in 2006 compared to 1996, the break-even effectiveness using the willingness-to-pay approach is a much lower 5.2 percent.

Summary

This analysis demonstrates that the choice of benefit-valuation method has a very strong influence on the outcome of a benefit-cost comparison. Even after accounting for the higher costs and decreased numbers of injuries and fatalities in 2006 compared to 1996, calculating benefits using a cost-of-illness method requires an effectiveness level that is more than six times greater than the effectiveness needed when using a willingness-to-pay method. Of course, OSHA's cost estimates were preliminary in 1998 and may have been underestimated. An improved benefit-cost analysis would need to reexamine the cost estimates. The benefit-cost comparison might be further improved by considering all injury types. Both OSHA and our analyses used only lost-workday cases as the basis for calculating the benefits of avoided injuries and illnesses, even though the benefit of a safety and health program is believed to apply to all types of injuries. Lost-workday injuries account for roughly half of total recordable injuries and illnesses (BLS, 2007). A more thorough benefit analysis that also included the benefits of avoiding injuries that are not in the lost-workday category would increase the benefit for a given level of effectiveness, decreasing the break-even effectiveness.

The 1998 analyses did raise several alternatives that might reduce the cost per injury prevented. One was to exempt small firms. Because of the element of fixed costs, small firms would tend to have higher costs per injury prevented. Critics of this exemption noted, correctly, that the weight of evidence indicated that small firms were more—not less—dangerous than larger firms (Mendeloff et al., 2006). They also noted that exemptions based on the number of employees had sometimes proved difficult to implement because the number of employees changes so frequently. A second strategy was to exempt firms in less hazardous industries. This approach has fewer implementation problems and serves to focus efforts where the risks are greater.

CHAPTER FIVE

Recommendations for Further Analysis

Our examination of the key concerns emerging from the 1998–1999 rulemaking process has revealed a number of uncertainties and limitations related to the anticipated effectiveness of safety and health programs, the implementation and enforcement of a safety and health program standard, and the benefit-cost trade-offs of a safety and health program standard. In this chapter, we present some recommendations for new analyses and other steps that would help address these questions and inform decisions about the merits of a safety and health program standard.

Effectiveness

Our examination of existing studies indicates that there is little evidence providing a clear indication about the effectiveness of mandatory safety and health programs. There is therefore a critical need for more detailed analyses related to the effectiveness of existing safety and health programs and related measures. Two types of analyses would help address the question of effectiveness.

Separate the Effect of Safety and Health Programs from Other Factors That Influence Injury Rates

More research focused on sorting out the relative contributions of various influences on changing injury and illness rates would help to isolate the effect of safety and health programs. Smitha, Kirk, et al. (2001) conducted such a study and made some valuable observations. At the same time, there are additional potentially important influences beyond those that they considered.

In particular, more research is needed to understand how changes in state workers' compensation programs could affect workers' reporting of injuries and illnesses. This would require going beyond the waiting-period and maximum-benefit variables and looking at other factors that affect the expected benefit to the worker of reporting. This could include caps on medical payments, limits on the duration of temporary disability payments, and other factors.

Examine in More Depth the Experience from Existing Programs

There is a need to look in more detail at the effect of safety and health programs on injury and illness rates. Most existing studies have examined effectiveness at the state level by comparing states with and without safety and health program standards. Such studies are confounded by differences across states in terms of workers' compensation laws, industry mixes, the safety and health programs themselves, and other factors. Conclusions from the few studies comparing participants to nonparticipants within states are suspect because voluntary participants are presumably more motivated, introducing a sampling bias. To overcome these problems, we suggest that it could be useful to compare injury rates in firms with safety and health program violations to size- and industry-matched firms without such violations. This approach would identify outcome differences between firms with and without functioning safety and health programs without the bias inherent in comparing participants and nonparticipants in a voluntary system or the confounding factors that come with comparing states with and without program standards. Specifically, we recommend carrying out the following tests:

- Examine whether establishments cited for safety and health program violations had higher injury rates prior to the citation than did other firms that were inspected but not cited. The reasoning here is that, if a safety program is effective in reducing injuries, firms with shortcomings in their programs should have poorer performance.
- Examine whether establishments cited for safety and health program violations show improvement in their injury rates subsequent to the citation compared to similar establishments that did not get cited. Here, one would be looking at the period *following* the citation. One might expect that firms respond to citations by making changes, since we know from follow-up inspections that firms usually do abate the violation (at least in the short term). If compliance with a safety and health program is important for injury prevention, we might expect to see improvements at these firms.
- Look for associations between safety and health program violations and intermediate metrics of effectiveness, such as measures of management commitment and worker engagement, changes in hazard-identification rates, and changes in violation rates of other OSHA standards. Though more difficult to measure, intermediate metrics could provide insight into the mechanisms by which safety and health programs might ultimately lead to safer workplaces. This type of information can be derived from "safety climate" surveys, which would ideally be administered prior to and after the introduction of the safety and health program mandate.

These analyses should be limited to inspected establishments, because there may be factors that are very difficult to observe that distinguish inspected from noninspected firms. The analyses could be enhanced by distinguishing cited establishments in terms of the type and number of separate sections of the safety and health program standard violated. Such analyses could also weight serious violations much more heavily than "other-than-serious" violations. This approach would provide insights about whether some elements of the safety and health program were more closely linked to performance than others. A limitation of relying on citation data is that enforcement policies and practices may vary by industry, by state, and over time. Inconsistencies in how citations and penalties are assessed would complicate these analyses. Here, we recommend examining some specific questions related to enforcement practices.

Implementation and Enforcement

Regardless of the question of the effectiveness of safety and health programs, acceptance of a mandatory standard will be impeded if employers are not confident that a standard can be clearly defined and fairly enforced. To help understand the validity of such concerns and how they can be alleviated, we recommend examining the issues associated with the implementation of state safety and health programs. It could be valuable to address the following questions:

- How frequently is the safety and health program standard cited relative to other standards, how often are such violations cited as "serious," and which elements of a safety and health program standard are most commonly cited? In California, for example, inspection data show that the safety and health program standard has been the most frequently cited standard in the state in most years since it was promulgated in 1991. These data also show that the violations were cited as "serious" in only about 10 percent of the cases. Such information could help employers comply with a new standard by characterizing the enforcement activity and highlighting which program elements may be the most difficult to implement and maintain. This information would also provide useful context for the effectiveness analyses proposed earlier: Situations in which safety and health program violations are rarely cited would not provide good tests of effectiveness.
- What are the states' enforcement policies, and are there relationships between different enforcement approaches and program effectiveness? States may differ in terms of their approaches to enforcing safety and health programs. Based on states' experience, it would be useful to identify policies associated with effective programs and consider whether they would provide a good model for OSHA's proposed standard.
- What type of training do inspectors receive to judge compliance and enforce the standard? Are there specific training tools or approaches that have been particularly successful?
- What sorts of communication efforts and other special assistance do states provide to employers prior to and during the early phases of implementation?
- What type of feedback have states received from employers regarding implementation and enforcement, and how have states responded to this feedback? In this case, it would be helpful to distinguish employers according to size and industry.

Benefits and Costs

Our analysis revealed several uncertainties related to benefit valuation and program costs stemming from assumptions or choices that were unclear or not well justified. Here, we outline some analysis and other steps that would help clarify some of these issues.

- Clarify the current industry baseline. OSHA's preliminary baseline estimate, based on data from the early 1990s, was that 51 percent of workers were employed at establishments that had compliant safety and health programs. The extent to which safety and health programs have become more or less prevalent since then is unknown. Given that the industry baseline is highly uncertain but undoubtedly still large, a clear understanding of this baseline is critical for accurately estimating benefits and costs of a safety and health program standard. The baseline could be estimated by distinguishing workers in three different categories of establishments: those with no safety and health program, those with a safety and health program that is less comprehensive than that in the proposed standard, and those with a safety and health program that is compliant with the proposed standard. Doing so would entail first characterizing mandatory state safety and health programs to determine the extent to which they match the proposed federal standard, as well as to which establishments the programs apply. (For example, programs in some states apply only to particular industries or to particular-sized establishments.) Full benefits and costs would then be assigned to workers at establishments with no program, no benefit or cost would be assigned to workers at establishments with compliant programs, and some degree of partial benefit and cost would be applied to workers at establishments with less comprehensive programs.

- Consider the impact of safety and health programs on all injury types. OSHA's existing analysis calculates this benefit by applying the anticipated effectiveness of safety and health programs to lost-workday injuries only. However, given the general nature of a safety and health program, there is no reason to expect that a program would preferentially reduce more severe injuries, and there is no evidence indicating this. Lost-workday injuries account for roughly half of total recordable injuries and illnesses (BLS, 2007). Thus, even though the benefit of avoiding a less severe injury is lower than that for a lost-workday injury, a more thorough benefit analysis that also considered injuries that are not in the lost-workday category would increase the benefit for a given level of effectiveness.

- Gather improved data on program costs. Cost data can be derived from interviews, site visits, surveys, and stakeholder input. While OSHA made a strong effort to accurately estimate costs, including holding several stakeholder meetings, in its rulemaking process, it is clear from the SBREFA process that more research would help produce more realistic and acceptable cost estimates. More input representing the diversity of work environments would provide insight into the real-world requirements and constraints of implementing a safety and health program, especially for small businesses. This is particularly important for developing more accurate and transparent estimates of the costs of controlling hazards identified through a safety and health program.

Bibliography

Bartis, James T., and Eric Landree, *Nanomaterials in the Workplace: Policy and Planning Workshop on Occupational Safety and Health*, Santa Monica, Calif.: RAND Corporation, CF-227-NIOSH, 2006. As of June 26, 2008:
http://www.rand.org/pubs/conf_proceedings/CF227/

Blincoe, Lawrence J., Angela G. Seay, Eduard Zaloshnja, Ted R. Miller, Eduardo O. Romano, Stephen Luchter, and Rebecca S. Spicer, *The Economic Impact of Motor Vehicle Crashes, 2000*, Washington, D.C.: National Highway Traffic Safety Administration, May 2002.

BLS—*see* Bureau of Labor Statistics.

Bureau of Labor Statistics, "Employment Cost Trends," Web page, undated. As of August 6, 2008:
http://www.bls.gov/ncs/ect/home.htm

———, "Industry Injury and Illness Data," Web page, updated October 16, 2007. As of June 26, 2008:
http://www.bls.gov/iif/oshsum.htm

———, "Census of Fatal Occupational Injuries (CFOI)—Current and Revised Data," Web page, updated April 17, 2008. As of June 26, 2008:
http://www.bls.gov/iif/oshcfoi1.htm

Burton, John F., Jr., "Workers' Compensation," in Kenneth G. Dau-Schmidt, Seth D. Harris, and Orly Lobel, eds., *Encyclopedia of Labor and Employment Law and Economics*, Cheltenham, UK: Edward Elgar, 2008.

Conway, Hugh, and Jens Svenson, "Occupational Injury and Illness Rates, 1992–96: Why They Fell," *Monthly Labor Review*, Vol. 121, No. 11, November 1998, pp. 36–58. As of June 26, 2008:
http://www.bls.gov/opub/mlr/1998/11/art3full.pdf

Forder, Paul K., and Robert D. McMurdo, *Working Together on Health and Safety: The Impact of Joint Health and Safety Committees on Health and Safety Trends in Ontario*, Ontario Workplace Health and Safety Agency, March 1994.

Hatch, L. L., P. G. Rentos, F. W. Godbey, and E. L. Schrems, *Self-Evaluation of Occupational Safety and Health Programs*, Cincinnati, Ohio: National Institute of Occupational Health and Safety, No. 78-187, October 1, 1978.

Hecker, Steven, Patricia A. Gwartney, and Amy E. Barlow, *Declining Occupational Injury Rates in Oregon*, Eugene, Oreg.: Labor Education and Research Center, University of Oregon, unpublished research, 1995.

House Committee on Small Business—*see* U.S. House of Representatives Committee on Small Business.

LaTourrette, Tom, D. J. Peterson, James T. Bartis, Brian A. Jackson, and Ari Houser, *Protecting Emergency Responders*, Vol. 2: *Community Views of Safety and Health Risks and Personal Protection Needs*, Santa Monica, Calif.: RAND Corporation, MR-1646-NIOSH, 2003. As of June 26, 2008:
http://www.rand.org/pubs/monograph_reports/MR1646/

Legislative Assembly of Ontario, Occupational Health and Safety Act (Bill 70), 1978, as amended through 2007.

Lewchuck, Wayne, A. Leslie Robb, and Vivienne Walters, "The Effectiveness of Bill 70 and Joint Health and Safety Committees in Reducing Injuries in the Workplace: The Case of Ontario," *Canadian Public Policy*, Vol. 22, No. 3, September 1996, pp. 225–243.

Mendeloff, John, *An Evaluation of the Maine "Top 200" Program*, prepared for the Occupational Health and Safety Administration Office of Statistics, 1995.

Mendeloff, John, Christopher Nelson, Kilkon Ko, and Amelia M. Haviland, *Small Businesses and Workplace Fatality Risk: An Exploratory Analysis*, Santa Monica, Calif.: RAND Corporation, TR-371-ICJ, 2006. As of June 26, 2008:
http://www.rand.org/pubs/technical_reports/TR371/

Morantz, Alison D., "Examining Regulatory Devolution from the Ground Up: A Comparison of State and Federal Enforcement of Construction Safety Regulations," Stanford, Calif.: Stanford Law School, Olin Working Paper No. 308, January 2007.

Occupational Safety and Health Administration, "Safety and Health Program Management Guidelines; Issuance of Voluntary Guidelines," *Federal Register*, Vol. 54, January 26, 1989, pp. 3904–3916. As of June 26, 2008:
http://www.osha.gov/pls/oshaweb/owadisp.show_document?p_id=12909&p_table=FEDERAL_REGISTER

———, Draft Proposed Safety and Health Program Rule, 29 CFR 1900.1, Docket No. S&H-0027, 1998a. As of June 26, 2008:
http://www.osha.gov/SLTC/safetyhealth/nshp.html

———, "Initial Regulatory Flexibility Analysis of OSHA's Draft Safety and Health Program Rule," 1998b.

———, "Preliminary Analysis of Effectiveness of Safety and Health Programs," 1998c.

———, "Preliminary Economic Analysis for Draft Safety and Health Program Rule," 1998d.

———, "OSHA's Mission," Web page, updated May 16, 2008. As of June 26, 2008:
http://www.osha.gov/oshinfo/mission.html

Office of Management and Budget, "Regulatory Analysis," Circular A-4, Washington, D.C., September 17, 2003. As of June 26, 2008:
http://www.whitehouse.gov/omb/circulars/a004/a-4.pdf

Ohsfeldt, Robert L., and Michael A. Morrisey, "Beer Taxes, Workers' Compensation, and Industrial Injury," *Review of Economics and Statistics*, Vol. 79, No. 1, February 1997, pp. 155–160.

OMB—*see* Office of Management and Budget.

OSHA—*see* Occupational Safety and Health Administration.

O'Toole, Michael F., "Successful Safety Committees: Participation Not Legislation," *Journal of Safety Research*, Vol. 30, No. 1, Spring 1999, pp. 39–65.

Policy Planning and Evaluation, Inc., *Regulatory Analysis of OSHA's Safety and Health Program Rule*, prepared on behalf of the U.S. Small Business Administration, Herndon, Va., 1999.

PPE—*see* Policy Planning and Evaluation, Inc.

Public Law 91-596, Occupational Safety and Health Act of 1970, December 29, 1970, as amended through January 1, 2004. As of June 26, 2008:
http://www.osha.gov/pls/oshaweb/owadisp.show_document?p_table=OSHACT&p_id=2743

Public Law 104-121, Small Business Regulatory Enforcement Fairness Act of 1996, March 29, 1996, as amended through May 25, 2007.

Reilly, Barry, Peirella Paci, and Peter Holl, "Union, Safety Committees and Workplace Injuries," *British Journal of Industrial Relations*, Vol. 33, No. 2, June 1995, pp. 276–288.

SBAR Panel—*see* Small Business Advocacy Review Panel.

Small Business Advocacy Review Panel, *Report of the Small Business Advocacy Review Panel on the Draft Safety and Health Program Rule*, Washington, D.C., December 18, 1998. As of June 26, 2008:
http://www.sba.gov/advo/is_repsh.html

Smitha, Matt W., Katharine A. Kirk, Kent R. Oestenstad, Kathleen C. Brown, and Seung-Dong Lee, "Effect of State Workplace Safety Laws on Occupational Injury Rates," *Journal of Occupational and Environmental Medicine*, Vol. 43, No. 12, December 2001, pp. 1001–1010.

Smitha, Matt W., Kent R. Oestenstad, and Kathleen C. Brown, "State Workers' Compensation: Reform and Workplace Safety Regulations," *Professional Safety*, December 2001, pp. 45–50.

Stanley, Marcus, *Essays in Program Evaluation*, dissertation, Cambridge, Mass.: Kennedy School of Government, Harvard University, November 2000.

Thomason, Terry, and John F. Burton, Jr., "The Effects of Changes in the Oregon Workers' Compensation Program on Employees' Benefits and Employers' Costs," *Workers' Compensation Policy Review*, Vol. 1, No. 4, July–August 2001, pp. 7–23.

Tolley, George, Donald Kenkel, and Robert Fabian, eds., *Valuing Health for Policy: An Economic Approach*, Chicago, Ill.: University of Chicago Press, 1994.

Tuohy, Carolyn, and Marcel Simard, "The Impact of Joint Health and Safety Committees in Ontario and Quebec," Toronto: Canadian Association of Administrators of Labor Law, January 1993.

U.S. House of Representatives Committee on Small Business, "OSHA's Draft Safety and Health Program Rule," hearing before the House of Representatives Committee on Small Business, 106th Congress, First Session, July 22, 1999.

Viscusi, W. Kip, and Joseph E. Aldy, "The Value of a Statistical Life: A Critical Review of Market Estimates Throughout the World," *Journal of Risk and Uncertainty*, Vol. 27, No. 1, August 2003, pp. 5–76.

Weil, David, "Are Mandated Health and Safety Committees Substitutes for or Supplements to Labor Unions?" *Industrial and Labor Relations Review*, Vol. 52, No. 3, April 1999, pp. 339–360.